Book 5
How Dense Are You?

Is this the right way to the moon?

Contents

Introduction	How Dense Are You?	3
Chapter 1	Weight and Mass	4
Chapter 2	How Far? How Full?	8
Chapter 3	How Dense?	16
Chapter 4	Technology Today — A Mirror on the Moon	20
Check Up	Alberta Einstein's Brain Tester	22
Don't Gloss Over the Glossary!		24
Indexed!		25

Alberta Einstein tries to measure the density of Eric Einstein's head. (The 'Huh' and 'Hmmm' routine)

Introduction

How Dense Are You?

You didn't really think we were talking about how intelligent you are did you?

We're talking about weighing and measuring - things like the mass, density and volume of blades of grass.

And gravity - the force that keeps your feet firmly on the ground.

Do You Know...
- how to measure mass and length?
- how to find volume?
- how weight and mass are related?
- how density is related to mass and volume?

Chapter 1   Weight and Mass

■ How are weight and mass different?

Everything is made of matter.

Look at the picture. All of the objects that you can see are made of matter. Even things that you cannot see, such as the gases in air, are made of matter. Matter is the "stuff" that all things are made of.

All matter is made of particles. The amount of matter in anything depends on the number of particles in it. The more particles, the greater the amount of matter.

Look at the balloons in the picture. Which balloon is filled with more matter?

What Is Gravity? Any two objects attract each other. Each object pulls the other toward its centre. This force that pulls objects toward each other is called **gravity**.

The balloon with more air has more matter.

All objects have a gravitational attraction for one another. The larger the object, the greater the attraction. The planet Earth is very large. Therefore it has a very strong gravitational pull on all objects near it. Since its pull is stronger than anything else near it, Earth pulls all objects toward its centre.

Gravity pulls everything toward Earth.

How does gravity change the astronaut's weight?

Gravity and Weight The amount of gravity pulling on any object depends on the amount of matter in the object. It also depends on the object's distance from the object that is pulling it.

Weight is a measure of the pull of gravity. The greater the pull, the more weight the object will have. The pull of Earth's gravity is greater on a bowling ball than it is on a golf ball. Objects that are pulled more have more weight. We say that objects that are pulled more by gravity are heavier.

Thank goodness we're a long way from Jupiter!

Eric Einstein feels the weight of the subject

TRY AND EXPLAIN THIS TO ERIC...

How can gravity be different on different planets?

THINK THINK THINK THINK THINK THINK THINK THINK THINK THINK

← Think Zone

Wonder what a mere mouse mass would be?

Weight and Mass How much do you weigh on Earth? On other planets, your weight would change. That is because the force of gravity is not the same on other planets as it is on Earth. For example, Jupiter's gravity is more than twice as strong as Earth's. Therefore, you would weigh twice as much on Jupiter as you weigh on Earth.

Although your weight changes from planet to planet, your mass remains the same. **Mass** is the amount of matter found in an object. It is not affected by the force of gravity.

The masses of small objects are measured in units called grams (g). A 10-cent piece has a mass of about five grams. The masses of larger objects are expressed in kilograms (kg). A kilogram equals 1,000 grams. Three small 300 mL cartons of milk have a mass of about one kilogram.

To find the mass of an object, put the object on one pan of a balance.

Add known masses to the other pan until the pans balance.

Mass is often measured in grams and kilograms.

6

Brain-teaser Time!

gravitational pull

a startled mass of elephant

perplexed scientist

weights

set of sillies playing about with balance pans

CRASH!

Experiment No 7642 crashes because of gravity.

Brainteaser: How would a change in gravity affect balance pans on Earth? Think of two other scenarios — one for a smaller planet and one for a much larger planet.

Chapter 2   How far? How full?

■ How can you find the length or volume of an object?

Have you ever watched the Olympics? If so, then you probably know something about measurement. Distances in all Olympic events are measured in metres.

Look at the picture. These swimmers are racing in an 800-metre event. The pool they race in is 50 metres long. In order to complete this race, they must swim 16 pool-lengths!

Help! I'm not King Kong save me!

15 lengths to go!

Distance   In some countries, distance is measured in inches, feet, yards and miles. These units are called imperial measurements. We use metric units to measure distances. Olympic distances are measured in units of metres. A doorknob is about one metre from the floor.

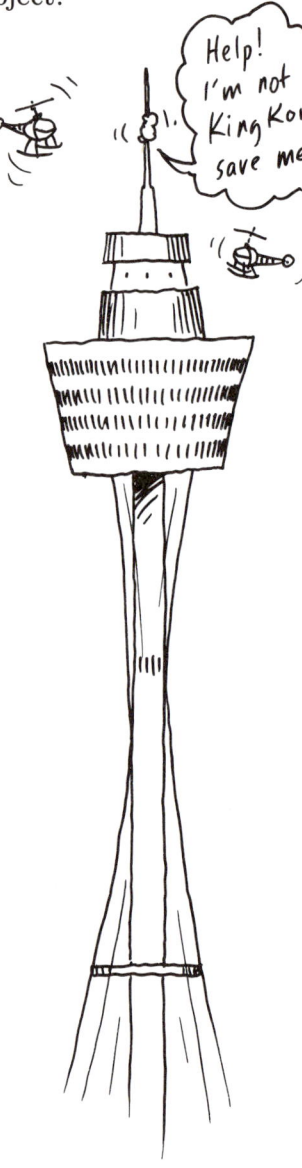

Eric Einstein measures Centrepoint Tower and his own bravery.

Many distances that you will measure are less than one metre. To measure such short distances as the length, width, or height of a shoe box, a smaller unit of length is needed.

If you look at a metre stick, you will notice that it is divided into 100 units. Each unit is called a centimetre (cm). A centimetre is 1/100 of a metre. The keys on most computer keyboards are about one centimetre wide.

Many distances are too long to be measured in centimetres or metres. They are measured in kilometres. A kilometre (km) is 1,000 metres. Look at the drawing below. The roadway between the bridge towers is about one kilometre long. Notice how many of each object would fit in this one-kilometre distance.

Measuring a centimetre

Cor! Is this a game of bridge?

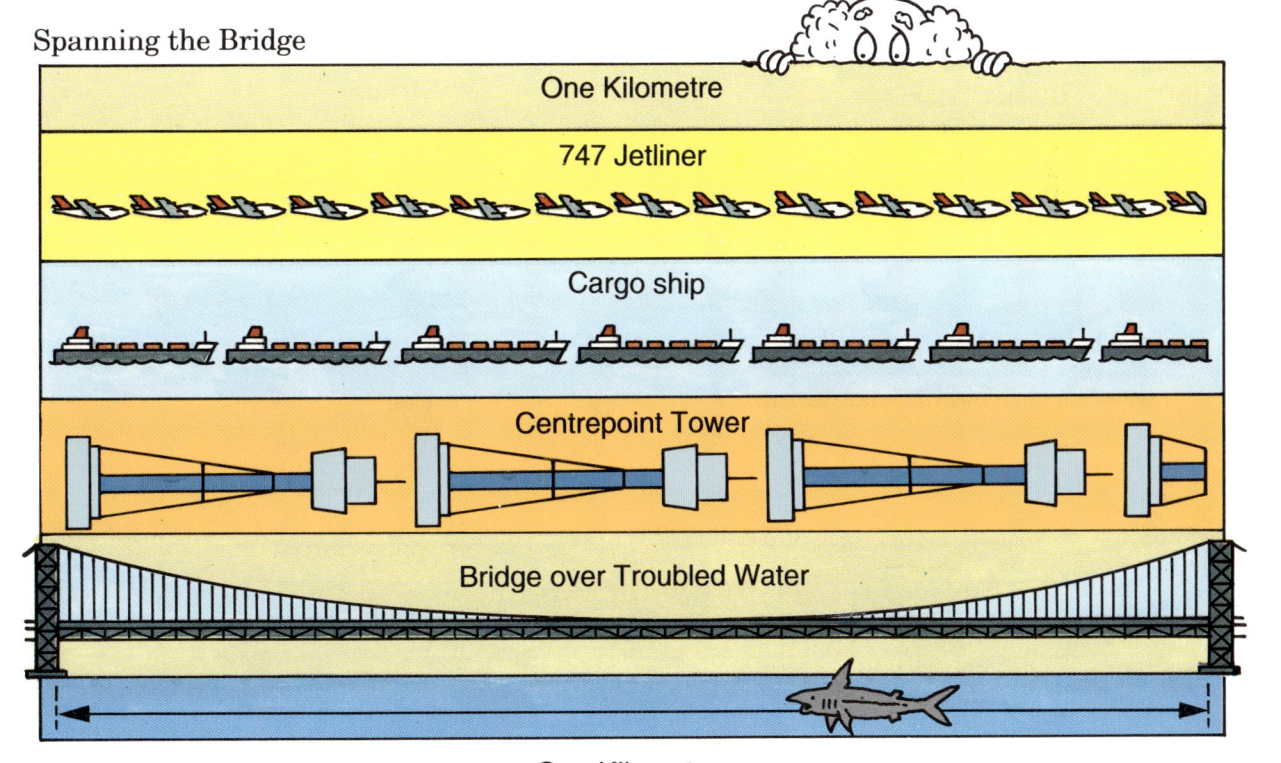

Spanning the Bridge

One Kilometre

747 Jetliner

Cargo ship

Centrepoint Tower

Bridge over Troubled Water

One Kilometre

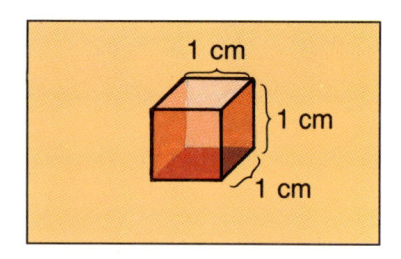

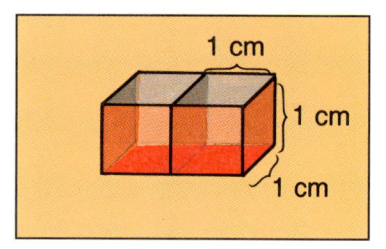

Double Volume

Volume Look around you. Every object you can see takes up space. You also take up space. The amount of space that anything takes up is called its **volume** (voll-yoom).

Look at the objects on the left. One object is made of one block. The other is made of two blocks. The object made of two blocks takes up twice as much space as the single block. Its volume is twice as large as the volume of the single block.

Look at the three groups of blocks at the bottom of the page. Each group has a different shape. But each group contains twelve blocks. Each group, therefore, takes up the same amount of space. Because each group takes up the same amount of space, all three groups have the same volume.

The Twelve Blocks Trick

Same space, same volume!

Measuring Volume The volume of liquids and solids can be measured. To measure the volume of a liquid, the liquid is poured into a special container that has a scale of measurement along its side. Measuring cups, graduated cylinders, and beakers like the ones below are used for liquid volume.

Measuring liquid volume with a graduated cylinder

Why are these containers good for measuring volume?

The volume of a liquid is measured in units called litres. A litre (L) of milk is a little more than four cups. To measure smaller volumes, we use smaller units. Look at the pictures of the graduated cylinders and beakers on this page. You will see that the scale on the side of each one of those containers is divided into millilitres (mL). A millilitre is 1/1,000 of a litre. That means that in one litre there are 1,000 millilitres. It takes about 15 drops of water to make one millilitre.

ALBERTA NEEDS YOUR HELP WITH THIS...
How would the scale on the side of a measuring container change if the container was fatter at the bottom than the top?

Think Zone

THINK THINK THINK THINK THINK THINK THINK THINK THINK

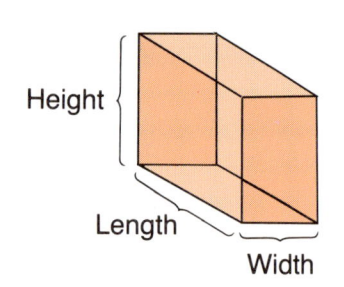

Volume = Length x Width x Height

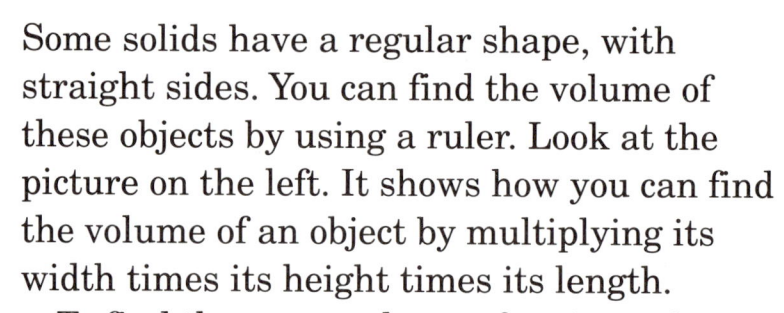

Some solids have a regular shape, with straight sides. You can find the volume of these objects by using a ruler. Look at the picture on the left. It shows how you can find the volume of an object by multiplying its width times its height times its length.

To find the exact volume of an irregular solid, you can lower the solid into a container of water. The solid will take the place of some of the water. Look at the drawing below. The rock takes the place of 20 millilitres of water. The volume of a solid can be measured in cubic centimetres. One cubic centimetre of a solid will take the place of one millilitre of water. The volume of the rock, therefore, is 20 cubic centimetres.

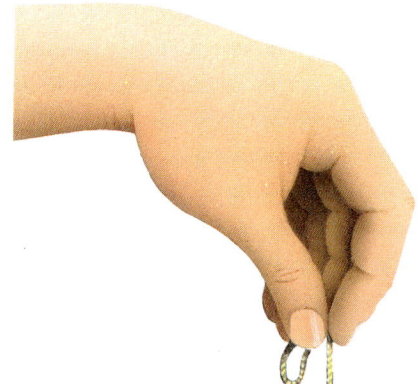

Clever Tricks!

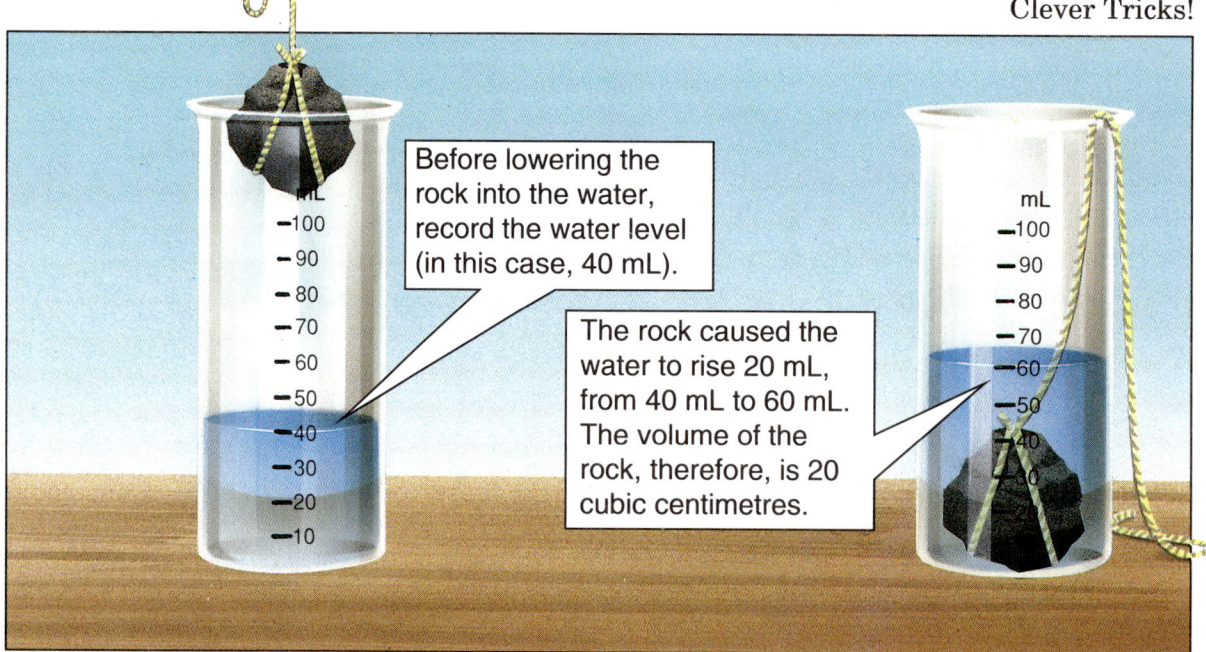

Before lowering the rock into the water, record the water level (in this case, 40 mL).

The rock caused the water to rise 20 mL, from 40 mL to 60 mL. The volume of the rock, therefore, is 20 cubic centimetres.

Measuring the Volume of an Irregular Solid

Maths Magic

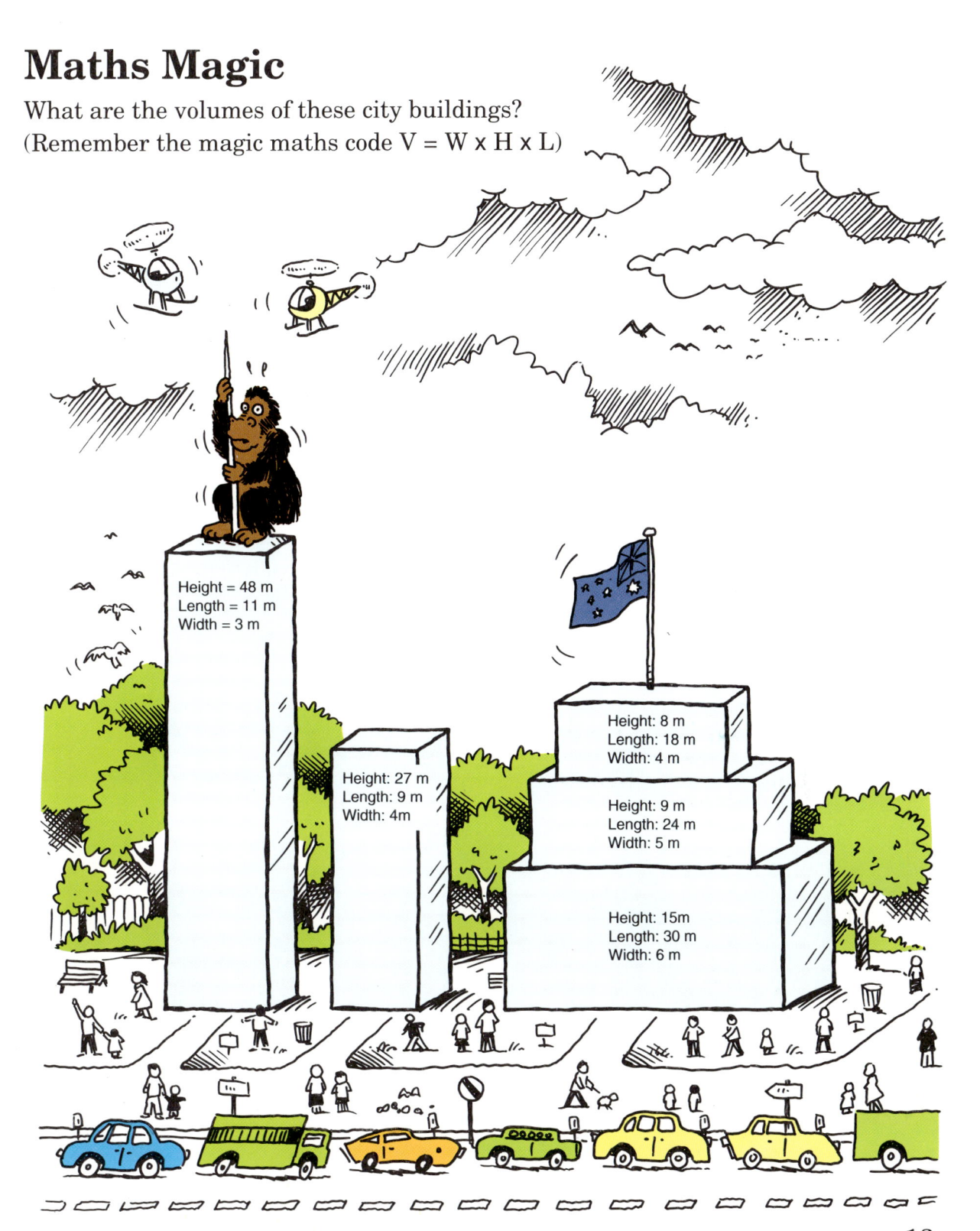

What are the volumes of these city buildings?
(Remember the magic maths code V = W x H x L)

Height = 48 m
Length = 11 m
Width = 3 m

Height: 27 m
Length: 9 m
Width: 4m

Height: 8 m
Length: 18 m
Width: 4 m

Height: 9 m
Length: 24 m
Width: 5 m

Height: 15m
Length: 30 m
Width: 6 m

Measure the Distance the Marathon Mouse has Run!

(But before you do, estimate how far you think it may be.)

start here

14

Alberta Einstein's Memory Tester

What is the volume of this ... er ...
... um ... irregular ... (that means 'odd shaped') ... solid? (Is Eric Einstein a solid?)
Experiment No. 8001. The Archimedes' Principle.

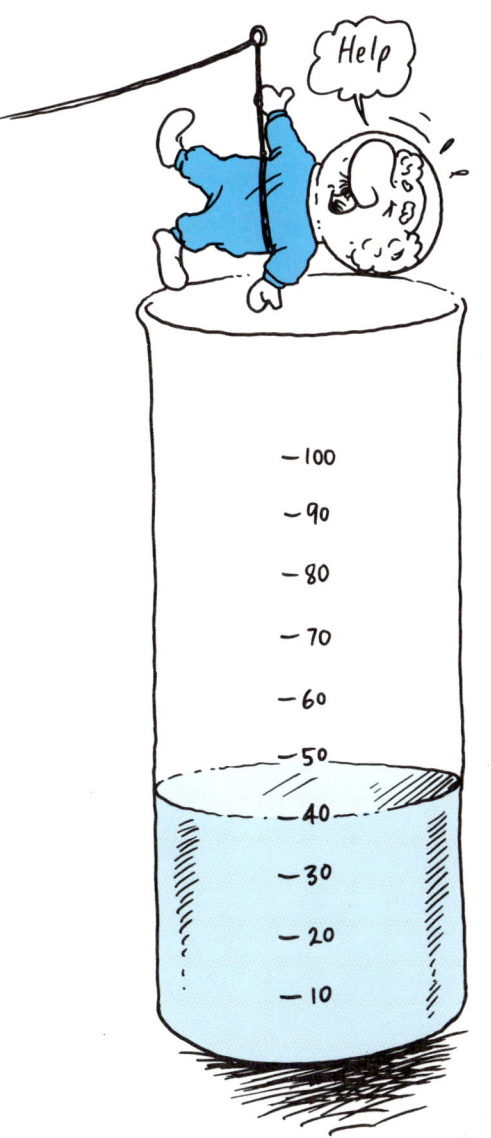

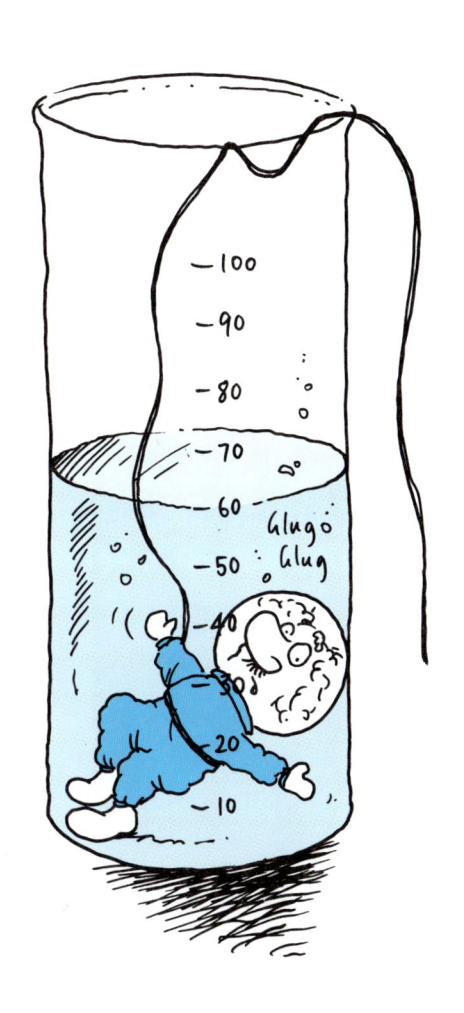

If you've forgotten how to solve this problem, turn back to page 12 ...

Chapter 3   How Dense!

■ How does density affect how heavy an object feels or whether it will float in water?

Objects that occupy the same amount of space often have different masses. The mass of an object depends upon the amount of matter making up the object. A bag of flour and a pillow have about the same volume. If you lift each of them, however, you will discover that the bag of flour has a greater mass.

Since the bag of flour has a greater mass, it must contain a greater amount of matter than the pillow. To fit within this volume, the matter within the bag of flour must be packed tightly together, or more densely. The pillow, however, contains less matter. To occupy this same volume, the pillow's matter must be spread out, or less dense.

Hey! Why's that kid staring at us?

Less

No overcrowding here

More

Packed tightly together — densely populated

What Is Density? Some kinds of matter have particles that are more closely packed together than particles of other kinds of matter. Therefore, different amounts of matter can fill the same amount of space.

If you picked up a bowling ball and a cloth ball of the same size, you would feel a difference in mass. The particles that make up the bowling ball are more closely packed.

Scientists use the term "density" to describe the way in which matter is packed together. **Density** is the amount of matter in a given space. The particles that make up a metal car are packed together more closely than the particles that make up a hot air balloon. For that reason, we say that a car is denser, or has a greater density, than a hot air balloon.

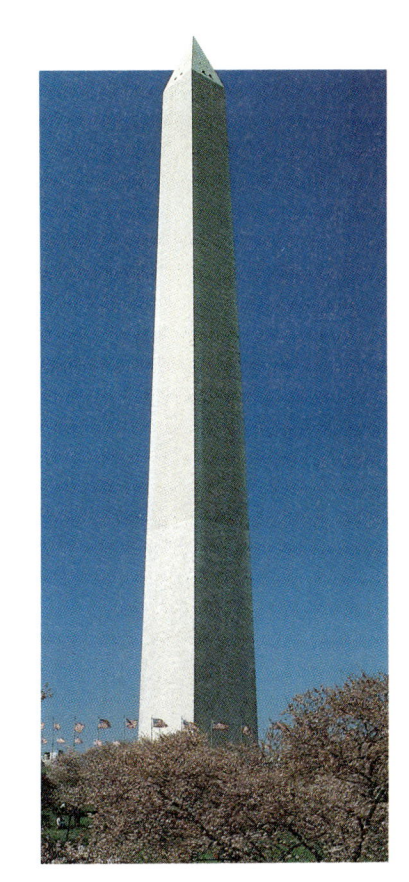

An object made of closely-packed particles

I wonder if I have a greater density than you.

There's only one way to find out!

What about us ping pong balls?

Which ball has a greater density?

'Song' of the solids ... the liquids ... the gases

Spot the different densities!

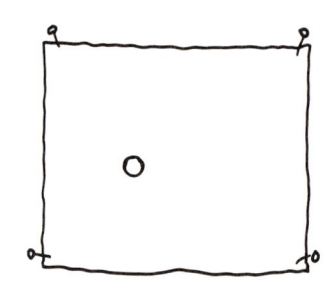

Name the density – solid, liquid or gas?

Cor!

Comparing Densities Usually, solids are denser than liquids, and liquids are denser than gases. Since the matter found in most gases is spread out, gases have a very low density. Solids are made of particles that are packed closely together. Solids, therefore, have a much greater density than gases.

Most liquids contain particles that are not as closely packed as those in solids. Liquids are therefore usually less dense than solids.

Density determines if an object will float. Objects less dense than water will float in water. Wood and oil are less dense than water, so they float in it. Objects more dense than water will sink. Iron is almost eight times more dense than water. When a piece of iron is placed in water, it sinks.

Brain-teaser Time

If iron and steel are *denser* than water, how can a ship that is made of iron and steel float?

19

Chapter 4   Technology Today

A Mirror on the Moon

When the astronauts landed on the moon, they left footprints there. They also left a special mirror on the moon. That mirror can be used to make measurements between two places on the earth's surface.

What kind of measuring device could stretch all the way to the moon and back? Certainly no ruler or measuring tape could reach that far. But a beam of special light, called a laser beam, can. Scientists know exactly how fast the light travels. They can calculate the distance to the moon from places on the earth.

Laser beams can also be used to measure other distances. They can tell how high clouds are above the earth. From an aeroplane, laser beams can even measure the height of steps in a stadium.

Think About It

Laser equipment is very expensive. Only specially trained people can use laser equipment for measuring. Laser measurements are very accurate. Do you think such accurate measurements are needed? Are they worth the expense? How could they make that much difference?

Mirror on moon helps to measure distance.

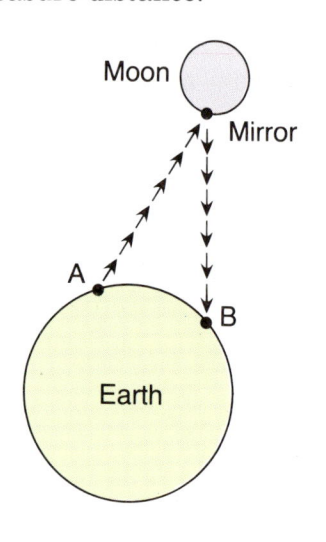

Measuring earth distance A to B with laser and moon

The Magic of the Moon

Since time began, people have marvelled at the moon. They've worshipped it, sung to it, woven stories around it and dreamed up ways to get to it.

Cyrano de Bergerac and Hans Pfaal flew hot-air balloons to the moon in the 17th and 19th centuries ... er, well, they did in their imagination!

Rub a dub dub A man in a floating tub!

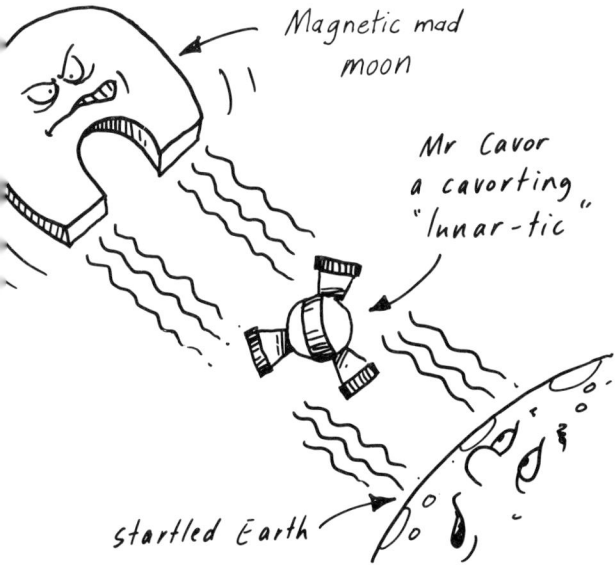

Magnetic mad moon

Mr Cavor a cavorting "lunar-tic"

startled Earth

One bright spark called Mr Cavor, designed a moon-bound craft in 1900. It was shaped like a ball, smeared with something magical called cavorite, lined with thick glass, and operated by shutters. When the shutters were raised, the anti-gravitational cavorite drew the craft to the moon — as if the moon were like a magnet!

The Baltimore Gun Club had an even better idea. Ninety-nine years before NASA got around to it, they designed a gigantic cannon to shoot a 'space-craft' of a very unusual kind to the moon. Three men and two dogs were on board. It took 97 hours, 13 minutes and 20 seconds to get within 40 km of the moon. Then it fell back to Earth and tumbled conveniently into the sea ... so they said ... but I think it's all moon-madness talk, don't you?

Hey! Did you read Jules Verne's book "From Earth to the moon"? It was written in 1865.

Wow!

21

Check Up   Alberta Einstein's Brain Tester

> **Summary**
> - Mass is measured in units called grams. Length can be measured in metres.
> - Volume is measured in litres or cubic centimetres.
> - Mass is the amount of matter an object contains. Weight is the measure of the pull of gravity on objects.
> - Density is the amount of matter in a given space.

Eric, you need a brain check-up! Heh! Heh!

Science Ideas
1. Make a list from **a** to **d**. Write the word in each list that does not belong.
   a. Some units to measure length are (metres, centimetres, litres).
   b. The volume of objects can be measured in (grams, litres, cubic centimetres).
   c. Mass can be measured in (grams, kilograms, kilometres).
   d. (Wood, Oil, Iron) has a density less than water, so it will float.

2. Make two columns on your paper, one for mass and one for weight. For each column, letter the items in order, from greatest to least. Then compare columns. What conclusion can you draw?

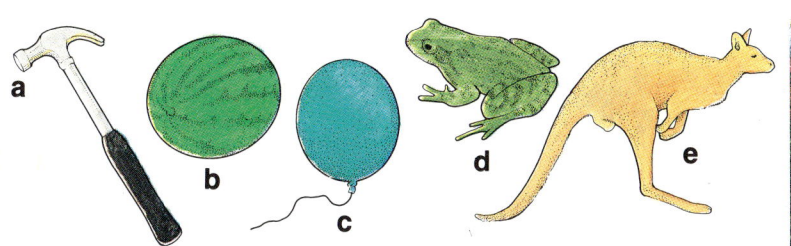

3. Look at the picture on the right. How does density affect what is happening in the picture?

4. **Data Bank**

   Use the table on the right to answer the following questions.
   1. If you weigh 45.5 kg on Earth, what would you weigh on Saturn?
   2. On which of the planets would you weigh the most? The least?

Weight on Other Planets	
Planet	Weight of object if 45.5 kg on Earth
Mercury	16.8 kg
Venus	40.0 kg
Earth	45.5 kg
Mars	16.8 kg
Jupiter	106.4 kg
Saturn	52.3 kg
Uranus	53.2 kg
Neptune	53.6 kg

5. **Problem Solving**

   As you jumped across a small pond on the way to school, some glass marbles and an apple dropped from your bag into the pond. How can you explain why the marbles fell to the bottom of the pond, while the apple bobbed about on top of the water?

23

Don't Gloss Over the Glossary!

anti — against
anti-gravitational — working against gravity/pulling *away* from the Earth
calculate — work out
cylinder — a mathematical shape (see diagram)
dense — thick
density — the amount of matter in a given space
device — an instrument or piece of equipment or apparatus
graduated cylinder — a cylinder with measurement markings on its side
gravitational pull — the force that pulls (see gravity)
gravity — the force that pulls objects towards each other
irregular shape — a shape that's not even/odd shaped
mass — the amount of matter found in an object
NASA — short for 'National Aeronautics and Space Administration'
regular shape — a shape with evenly matched sides

I feel kind of dragged down!

Jupiter

Eric feels the force of Jupiter's gravitational pull.

stadium — a large building that seats enormous crowds of people who come to watch events.
volume — the amount of space anything takes up
weight — a measure of the pull of gravity